BODY TALK
MOVEMENT
THE MUSCULAR AND SKELETAL SYSTEM

JENNY BRYAN

Wayland

BODY TALK

BREATHING

DIGESTION

MIND AND MATTER

MOVEMENT

REPRODUCTION

SOUND AND VISION

SMELL, TASTE AND TOUCH

THE PULSE OF LIFE

Editor: Catherine Baxter
Series Design: Loraine Hayes
Consultant: Dr Tony Smith – Associate Editor of the *British Medical Journal*
Cover and title page: British athlete Fiona May in action.

First published in 1992 by Wayland (Publishers) Ltd.
61, Western Road, Hove, BN3 1JD, England

British Library Cataloguing in Publication Data
Bryan, Jenny Movement. – (Body Talk Series)
I. Title II. Series 612.7

ISBN/07502 0414 1

Typeset by Key Origination, 1 Commercial Road, Eastbourne
Printed in Italy by G. Canale & C.S.p.A., Turin
Bound in France by A.G.M.

CONTENTS

Introduction   4

Standing up   6

Walking   8

Running   10

Lifting   13

Weightlessness   16

Co-ordination   18

Reflex movements   21

Slowing down   24

Immobility   27

Replacing limbs   29

Anaesthetics   32

Sports injuries   34

Avoiding injury   36

Full speed ahead   38

Body language   42

Staying mobile   44

Glossary   46

Books to read   47

Index   48

INTRODUCTION

It is impossible to keep completely still! Even when you are asleep part of your body is moving. Each time you move, your nerves, muscles, bones, joints, tendons and ligaments must work together so that your body does what you want it to do. Your nerves are the messengers. They carry instructions to your muscles which tell them when to contract and when to relax.

There are two main types of muscle in the body: smooth muscle and skeletal muscle. You never know when your smooth muscles are working because they are deep inside your body - in your intestines, your lungs and your arteries. Your heart has its own special type of muscle that is different from both skeletal and smooth muscle.

The muscles you can feel or see are skeletal muscles and you can control how they work. Each of these muscles is attached to two or more bones of your skeleton, that's why they are called 'skeletal'. When a muscle contracts it gets much shorter and it pulls the two bones it is attached to closer together. That is how you move.

Muscles are attached to bones by tendons and bones are linked together by joints. These include hips, knees and elbows. All our movements would be very stiff if we didn't have joints! Try walking without bending your knees!

Joints enable us to bend our limbs but they don't hold our bones together. That is the job of the ligaments, which make the connection between the bones in a joint. They are the binding ropes of the skeleton.

Your nerves, muscles, bones, tendons and ligaments must all be in good working order for you to be able to move smoothly and easily. Some movements only need a few muscles, others – such as running – need great co-ordination of many muscles all over your body.

As we get older we gradually lose some of our mobility (the ability to move around). Joints wear out, bones become fragile, muscles become weak. But, by keeping fit when we are young, we can strengthen our bodies and help ourselves to stay mobile when we are older.

LEFT Try as you might, you'll never be able to stand as still as these statues.

OPPOSITE ABOVE Which parts of their bodies are these children moving as they play in the water? Why do you think they are moving their arms as well as their legs?

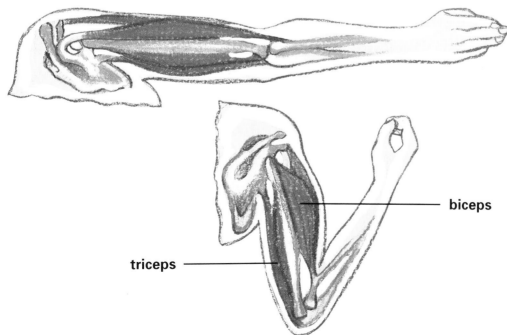

LEFT When you bend your arm, the muscles contract. As a muscle contracts it becomes shorter. This means that the bones in the limb that you want to move are pulled closer together.

biceps

triceps

STANDING UP

It takes at least six sets of muscles, forty bones and the same number of joints to stand up, and far more to keep you upright. No wonder it takes most babies a year to get to their feet and another year to stop falling over a lot of the time!

When you push yourself out of your chair your feet act as the launch pad for your upward movement. The bones in your feet are called metatarsals and your toebones are the phalanges.

The muscles in your bottom make you go forwards and the muscles in your thighs allow you first to bend and then to straighten your legs. Your hamstring muscles make sure that your hip and knee joints move too.

Toddlers often fall over when the messages to their brains are faulty.

Once you're on your feet the calf muscles in the lower part of your legs take the strain. The three bones in each of your legs support your body and act as pillars. Between your knee and ankle, two bones run alongside each other. The larger one is the tibia (shin bone) and the smaller one is the fibula. The bones in your ankles and heels are called the tarsals.

From hip to knee runs the biggest bone in your body. It is called the femur. The femur slots into your pelvis (hip bone) which, in turn, is linked to your spine (back bone).

The spine carries the weight of the body and supports the head. It consists of over thirty bones called vertebrae. They keep the back flexible; they aren't stuck together like a rod. Cushions of tissue, called discs, lie between the vertebrae. They act as shock absorbers and make sure that the spine can bend in all directions – forwards and backwards and slightly sideways.

Running down the centre of your spine is the spinal cord - the nerves that tell the rest of your body what to do. When you want to stand up your brain sends messages down these nerves to the muscles in your legs and feet. It takes a fraction of a second for those instructions to be received and understood. By then, you're ready to go!

BALANCE

You wouldn't be able to stand upright for very long without a system to help you keep your balance. Deep inside each ear you have a series of semi-circular canals, lying at right angles to one another. As you move your head the fluid inside these canals also moves. Nerves record these movements and tell your brain where your head is in relation to the things around you. The brain interprets the information and instructs other parts of your body to move accordingly, so that you keep your balance.

The skeleton

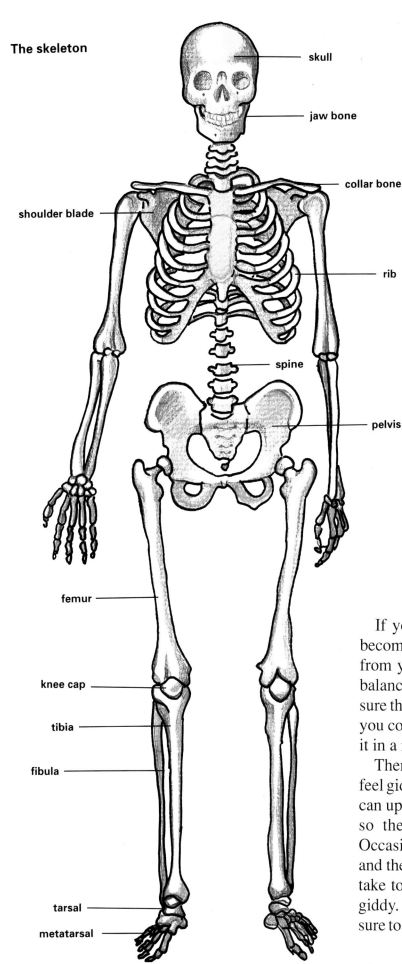

- skull
- jaw bone
- collar bone
- shoulder blade
- rib
- spine
- pelvis
- femur
- knee cap
- tibia
- fibula
- tarsal
- metatarsal

ABOVE This dancer has a highly-trained sense of balance which allows her to spin round and round and then leap across the stage. She has been practising for a long time. If you tried to do the same thing you would feel dizzy and you might fall over.

If you spin round in a circle your brain may become so confused by the signals it is receiving from your ears and eyes that you will lose your balance and fall over. If you want to try this, make sure that there aren't any sharp objects around that you could fall against. It is also a good idea to try it in a room with a nice soft carpet!

There are many other reasons why people may feel giddy and fall over. An infection inside the ear can upset the delicate fluid balance in the canals, so the wrong signals are sent to the brain. Occasionally, there is too much fluid in the canals and the signals get jumbled. Or drugs that people take to treat other illnesses may make them feel giddy. If you keep feeling giddy you should be sure to tell your doctor in case you need treatment.

WALKING

Each time you put one foot in front of the other some muscles in your legs contract while others relax. When the muscles that make the leg bend (flexor muscles) are contracting, those that make it straighten (extensor muscles) are relaxed.

Take a step forward with your right foot. Do it in slow motion and watch what happens. First, your calf muscle contracts and lifts up your right heel. This presses the ball of your right foot into the ground and makes your right knee bend slightly.

Can you feel your weight shifting? At this stage, it's almost impossible not to move forwards. As you straighten your right leg, your heel is the first part of your foot to touch the ground. At the same time the muscle in your left calf is contracting and your left heel is leaving the ground.

But how do your muscles know when to contract? They depend on messages from your brain. These messages are sent down the long nerves in the spinal cord as electrical signals. The signal is produced when charged molecules move backwards and forwards across the membrane that protects the nerve.

Chemicals called nerve transmitters 'carry' the electrical signals from nerve to nerve and from nerves to muscles. As soon as a muscle receives a signal, it contracts. Between signals it relaxes.

Muscle is made up of muscle cells, or fibres. They are cylindrical and lie alongside each other. Each fibre contains a group of fibrils. When you see these under a microscope they look striped. This is because each fibril contains two proteins, actin and myosin. Actin is pale and myosin is dark. When the actin and myosin proteins slide over one another the muscle contracts.

A muscle needs energy to contract. It gets this by breaking down carbohydrate from food. If there is no carbohydrate around, it can use fat.

Muscles are very energy efficient. When you are walking around you use only a few calories.

These feet were made for walking – and that's just what they are doing! Some people enjoy leisurely country walks, others take part in walking races. They can walk nearly as fast as someone who is running slowly.

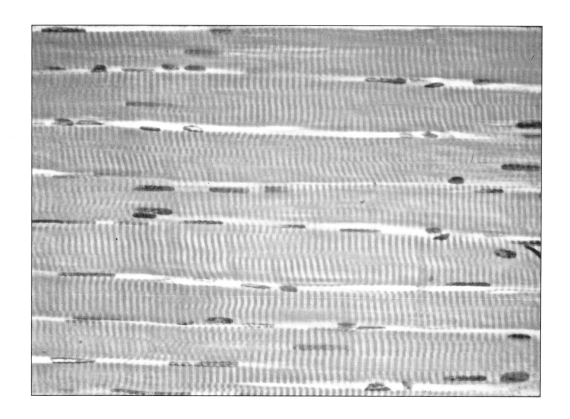

LEFT Under a microscope you can see the small pink fibres that make up muscle. They contain groups of even smaller fibrils which lie parallel to each other.

BELOW How on earth does this contortionist do it? He is bending his arms and legs in a way they weren't designed for. Let's hope he doesn't pay for it later with damaged joints, pulled muscles and torn ligaments!

INSIDE A JOINT

Some joints are more mobile than others. At one extreme, the bones of the skull fit firmly together with no movement allowed. But at the other extreme, the hip and shoulder have ball and socket joints which allow the bones to move in most directions. For example, your hips allow you to bend forwards or backwards, to twist and turn your upper body and move your legs outwards.

The elbows and knees are hinge joints. So, like a door, they can only move backwards and forwards. Saddle joints, like those at the bottom of the fingers, allow the bones to move sideways as well as backwards and forwards. Wrist movement is similar but more restricted, because the wrist bones are joined by gliding joints.

Inside a moving joint there is a gap between the two bones. If the bones rubbed together it would be very painful. The gap contains lubricating fluid.

The tip of each bone also has a protective layer of cartilage (gristle) which is tough and strong. Like the discs in the spine, cartilage acts as a shock absorber for the joint. Without it, you could injure yourself doing simple things like walking down the stairs.

RUNNING

Are you a sprinter or a long-distance runner? If you're a sprinter you'll need to be able to run in high-powered bursts. But if you're a future marathon runner you'll need stamina.

The type of running you're best at will depend on the sort of muscles you were born with. Top sprinters are born with mainly 'fast muscle fibres'. These take less than a tenth of a second to contract! Long-distance runners have mainly 'slow muscle fibres', which take three times as long as fast fibres to contract.

All muscles - fast and slow - need energy to contract. They get their energy from carbohydrate in our food. The body breaks this down to glucose, which is stored in muscle as glycogen.

Slow muscle fibres need plenty of oxygen to convert glycogen into energy. But fast muscle fibres don't need oxygen to process glycogen, they can get energy from it by a more direct route. This enables them to contract more quickly, but the drawback is that they get tired sooner than slow muscle fibres.

ABOVE Linford Christie (centre) bursts from the blocks. OPPOSITE The New York marathon.

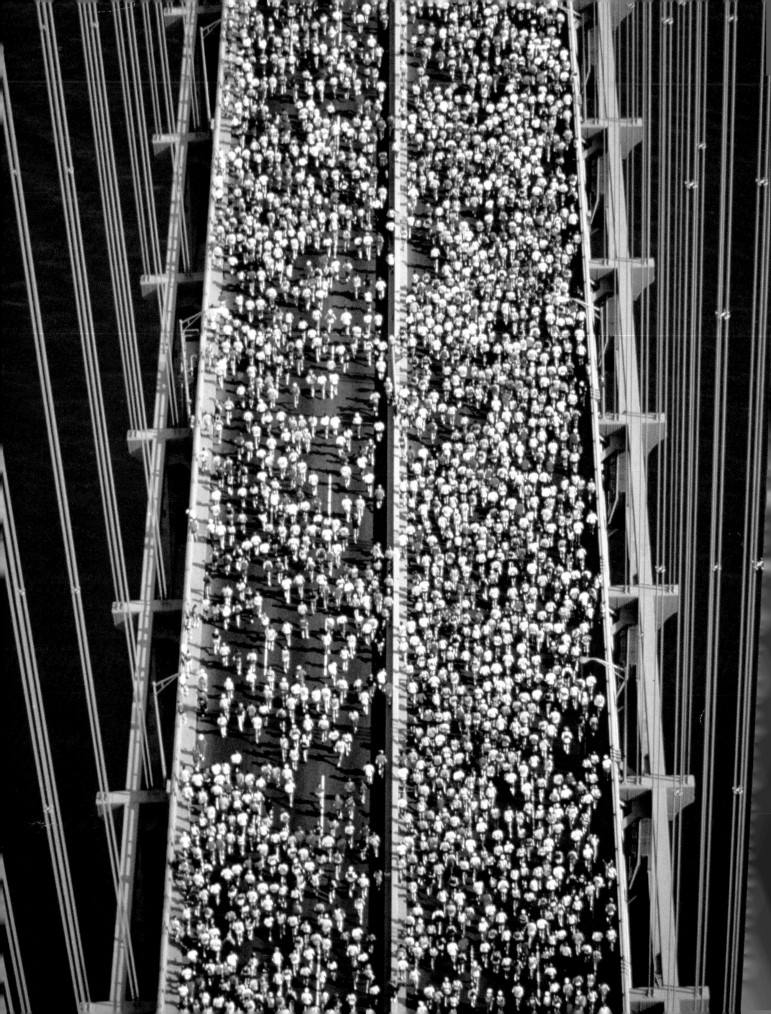

If you're preparing for a marathon these are the sorts of foods you need to eat. They provide you with the carbohydrate that your body needs to store as glycogen ready to produce energy.

People with mainly fast muscle fibres are best at sports that require short, sharp bursts of activity, like sprinting, throwing and jumping. But people with mainly slow muscle fibres are better suited to stamina sports such as long-distance running.

Most of us have roughly equal numbers of fast and slow muscle fibres. We cannot hope to be outstanding at speed or stamina events. Even long-distance runners need some fast fibres to enable them to sprint the last 50 metres to the finishing line. They must judge the race very carefully. If they race for home too soon their fast muscle fibres will tire before they reach the line and they will grind to a halt. But if they leave it too late, they'll lose their chance of a medal.

The best food to run on is carbohydrate – foods like pasta, rice and bread. When the glycogen supply from carbohydrate has run out the body must use fat for energy. But this takes a lot more oxygen, and a marathon runner may not have a lot of oxygen to spare at the end of a long race.

Will-power is important if the runner is to produce a good performance. Runners can train hard and have the best equipment, but without the will to win they won't pass the finishing line.

CRAMP

Even the finest athletes sometimes have to drop out of vital races because of cramp. Cramp is the painful and disabling result of a build up of lactic acid in the muscles. It happens when the muscles don't have enough oxygen to metabolize glycogen and produce energy.

Cramp can, of course, occur when you sit or lie down awkwardly and restrict the blood flow to part of your body. Again, this means that your muscles don't get enough oxygen so they go into spasm.

LIFTING

How strong are you? For most sports you need to be fit and, for some, you need stamina. But you need strength to be able to lift heavy weights.

Your muscles use different kinds of actions to carry out different tasks.

When you stand up, walk or run, your muscles contract isotonically. This means that all the effort goes into moving parts of your body. When you hold a heavy object, your muscles contract isometrically. Instead of moving parts of your body, the effort goes into opposing the natural gravitational forces pulling the object to the ground. During a normal day a muscle may be working isotonically one minute and isometrically the next.

Look's like hard work doesn't it? Are this weightlifter's muscles working isotonically or isometrically?

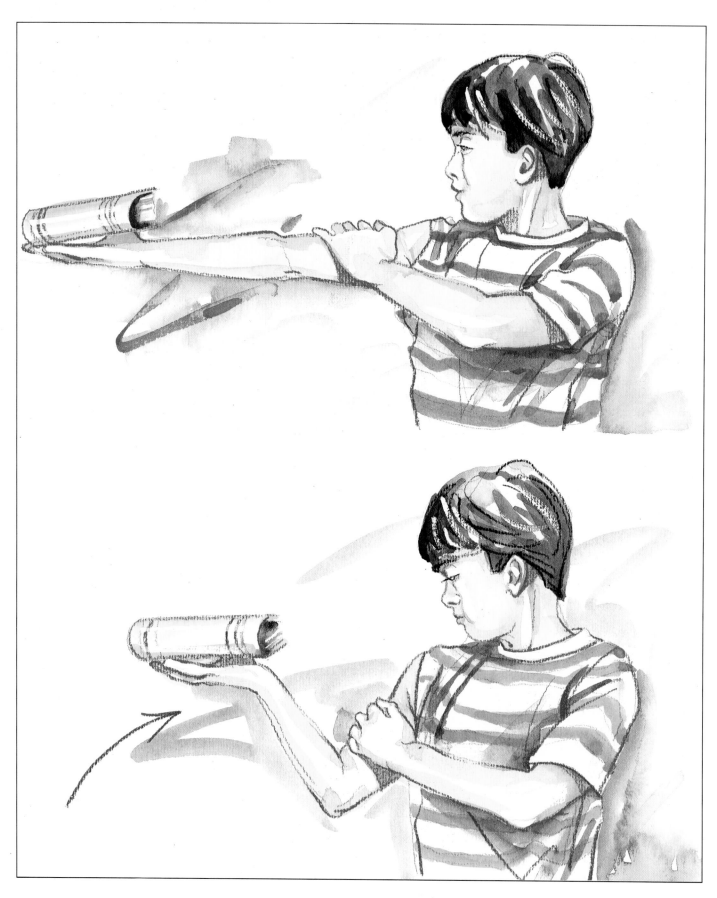

This boy can feel his biceps shortening as he lifts the book.

Try this. Pick up a heavy book and support it in your outstretched hand. Feel the biceps in that arm – they are the muscles at the front of your upper arm. They feel taut with the effort of supporting the book and keeping your elbow straight, but they have not shortened. They are acting isometrically.

Now, bring the book up towards your shoulder. This means you have to bend your elbow. Keep your other hand on your biceps and you will feel them shorten as your arm moves. Now, your muscles are acting isotonically.

You can train your muscles to become bigger and stronger. A planned programme of exercise can increase the size of muscles by up to 60 per cent. Someone who is very strong will have much bigger muscle cells with more fibrils. Most people are at their strongest by the time they are twenty and stay strong until about forty. After that, people gradually lose strength. By the age of sixty, the maximum amount of work their muscles can do may be only about half of what they could do when they were young adults.

LIFTING THINGS PROPERLY

It is very important to lift heavy things correctly. If you don't, you may damage your back. Never bend from the waist or keep your legs straight when you lift something from a low position. You will be putting too much strain on the vertebrae and discs in your spine.

Always stand as close as possible to the thing you want to lift. Then bend your knees so that you go down to the same level as the object. Take a firm grip on it and straighten your legs so that you lift it up. Keep your back straight.

WEIGHTLESSNESS

When astronauts set foot on Earth after several months in space, some can barely stand. That is because they have been weightless during their space flight. In space there is no gravity to pull you towards the ground, so you float.

If an astronaut wants to move, he or she gently pushes forward. It takes very little effort and the muscles do not need to work hard. After some time of this easy action, it's quite a shock when the muscles experience normal gravity again.

Bones get an even greater shock! The bones of people who spend long periods in space get lighter and thinner. Since they aren't having to support the astronaut's full weight, the rate at which new bone is formed gradually reduces.

Bone may look dead and inert, but it's just as alive as the rest of your body. It is made up of bone cells embedded in a firm matrix which acts as a depot for the body's calcium. The bone cells are constantly dying and being replaced. In a person up to the age of twenty, more cells are made than die. So the bones grow and are reshaped. Over the next twenty years, the bones get stronger and more solid and the skeleton reaches peak strength.

If someone is weightless for long periods, it isn't necessary to build up the strength of the bones. So the body doesn't bother to replace the bone cells that die, and the bones get weaker. As long as an astronaut stays in space, it probably doesn't matter that their bones are weaker. But coming back down to Earth, their bones have to take the full weight of their bodies again, and they may break. Only a few astronauts have ever stayed in space long enough for their bones to get thinner and, when they came back to Earth, they were able to make up the loss by making more bone.

If long space flights become routine, in the next century, scientists will have to find ways of combatting bone loss in space, because there comes a point when the damage cannot be repaired.

WATER THERAPY

Being weightless for short periods can be good for you. Taking exercise in water helps to get stiff joints moving and strengthen flabby muscles. That is why many people with bone, joint and muscle problems have 'hydrotherapy'.

In water they can gently exercise parts of their bodies which are too weak to move far on dry land. The water supports the body and takes some of the weight off the injured limb.

Hydrotherapy also boosts people's confidence. As their muscles get stronger and their joints more mobile, they feel encouraged to move around when they leave the comfort of the swimming pool.

It may look easy, but floating in space takes some doing. You must learn when to hold on to things to stop you colliding with equipment.

CO-ORDINATION

Can you remember learning to write? It took several months just to learn the basics and your handwriting is probably still changing and getting better. Writing is one of the most complicated movements that we make. It needs great co-ordination between eyes and hands. Your eyes and brain recognize the shapes of letters and the combinations of them that make words, and the muscles of your hand have to receive the right signals from them to form the writing on the page.

Co-ordination is something which comes gradually. When it is born, a child has all the bones, joints, muscles and nerves that it needs to walk, run, play and write. But it does not know how to use them and combine their activities effectively. In fact, it has no co-ordination.

Slowly, a baby becomes aware of its hands. First it looks at them and sucks them. By about four months old it can hold things that it is given, but it can't pick them up. It starts to pull at its clothes and to co-ordinate eye and arm movements. After another month, it can pick things up, clumsily at first and then more carefully.

At one year old, a child can pick things up between thumb and forefinger - something which takes precise muscle control. He or she will start fitting things together, such as simple games or puzzles, and moving toys around.

At two, a child can hold a pencil, but cannot control it. At first there will be big scrawls. Very slowly, the movements will get smaller and more precise. But it's only once a child is shown how to make the shapes of different letters that he or she will begin to control the pencil. Some letters (such as 's') are more complicated than others (such as 'i') and require greater co-ordination.

Carefully does it! Look at the concentration on this toddler's face. You could put the ball on the stick with no trouble at all. But he must learn how to co-ordinate his movements so the stick goes through the hole.

Some children are quicker at learning to write than others.

All children learn the fine movements that are needed for writing at their own speed. They will need the same sort of movements for other activities that require precision, such as painting, sewing and woodwork. Some learn faster than others. A few parents try to give their children a head start in learning to read and write by showing them cards with letters on them when they are only a few months old. They may even get them to read and write when they are only two or three years old – long before they go to school.

This process of teaching children basic skills while they are very young is called 'hot housing'. Nobody knows if this will help the children to do better in later life. They may be ahead when they first go to school, but do they stay ahead? Or will they lose interest in lessons while they wait for the other children in their class to catch up?

Your physical co-ordination affects all kinds of movements and activities. Some children are naturally rather clumsy. Because of poor co-ordination between their eyes and bodies, they tend to bump into things or knock them over. They aren't very good at ball games that require co-ordination. Clumsy children are just as clever as other children and they can find other outlets for their skills. For example, if ball control is a problem they may prefer swimming, running, dancing or weight training.

RIGHT Scoring a goal can be harder than it looks. Your brain must process a huge amount of information.

BELOW Twenty years ago this child might have been forced to use his right hand to write. Luckily, left-handers today are allowed to do what comes naturally.

LEFT- OR RIGHT-HANDED?

Most people are right-handed. But about one in ten men and slightly fewer women are left-handed. A few people can use both their left and right hands equally well.

Which hand you prefer to use depends on how your brain has developed. The brain is wired up so that the nerves that control the right side of your body start in the left side of your brain, and those that control your left side start in the right side of your brain. A person who is right-handed is probably more developed in the left side of the brain and someone who is left-handed is probably more developed on the right side.

Most things in life are geared to the needs of right-handed people just because there are more of them. But a left-handed child should never be forced to write with his or her right hand, as this may cause severe and unnecessary learning problems.

REFLEX MOVEMENTS

If you go to the doctor for a check up, he or she will probably do a simple test to see if your nervous system is working normally. It's called the knee jerk reflex. You can try it out yourself.

Cross your legs so that one foot is off the floor. Give your knee a sharp tap just below the knee cap. The lower part of your leg will swing upwards. (It isn't easy to test your own reflexes, so if your leg doesn't swing up get someone else to tap your knee – make sure you don't kick them!)

How does this happen? When you tap your knee you stimulate the tendon of the quadriceps muscle, which is attached to the bone in your knee. The muscle stretches and sends messages back up the nerves from the muscle to your spinal cord. Instead of going all the way back to the brain to be processed, the information is passed to a nerve that takes it straight back to the thigh muscle, which contracts so that your leg swings up.

Reflexes are thus short cuts for the nervous system. They are very useful because they allow you to react very quickly in dangerous situations. For example, if you touch something hot you automatically take your hand away. It takes a fraction of a second. If your brain had to process the information before you could withdraw your hand, it would take much longer and you would be badly burned.

We are born with this sort of reflex. But scientists have proved that some reflexes can be learned. In the early 1900s a Russian scientist called Ivan Pavlov showed that it was possible to train dogs to salivate when he rang a bell. We all salivate when we smell food - we only have to think about our favourite foods and saliva comes into our mouths! But Pavlov began by giving his dogs food when he rang a bell. When they got used to that, they started to salivate when they heard the bell.

THE KNEE JERK REFLEX

This is a reflex test which you can try out for yourself or with the help of a friend:

1) Sit down and cross your legs like the girl in the picture.

2) Using the side of your hand give your knee a sharp tap.

3) If at first your leg doesn't swing up, try again – or get a friend to help you. You may not be tapping the muscle correctly.

Sportsmen and women use similar methods to improve their skills. For example, a top-class tennis player who stands at the net will volley instinctively when a ball comes towards him or her. He or she does not have to think about the necessary movements – the body does them automatically because they have been practised so many times. At the beginning of the training routine, however, he or she would have had to learn the right ways to reach and hit the ball.

Another winning volley from tennis champion, Steffi Graf. Even top level players have to use immense concentration to get it right.

Does your mouth water when you look at this picture? Most of us salivate when we see, smell or even think about our favourite food. It's a reflex action that we have no control over.

TWITCHES AND TREMORS

Some people move without meaning to. This is called a twitch or tremor. If you hold your arms out straight you will see a very slight tremor in your fingers. This is normal.

It is not normal to have a frequent, noticeable tremor in your hands, feet or face. This occurs when the brain is sending too many signals to the muscles. It can happen as we get older and the chemicals in our brains stop working properly. Or it may be caused by injury to the brain after an accident. Some drugs, including alcohol, can have the same effect, which may become permanent if you keep taking them. Even drinking too much coffee can make your hands shake. Although this usually wears off quite quickly, if you want to avoid the problem in the first place, (and get a better night's sleep), switch to decaffeinated coffee!

Don't confuse nervous twitches and tremors with shivering. Shivering is one of the body's natural protective mechanisms against feeling cold. Your muscles contract by themselves in an effort to produce heat.

Shivering helps to make heat so that you warm up quickly.

SLOWING DOWN

As people get older, many of the control systems for the way the body moves start to slow down. In middle age, people cannot run as fast as they used to, and their limbs are less flexible. With age, joints get stiff and bones become thin.

It's a vicious circle. As people start to slow down, they move around less and so their muscles become weaker. This makes them even less willing to take some exercise and their joints, bones and muscles become even weaker.

Joint stiffness is usually due to arthritis. That means inflammation (swelling) in the joints. But, strictly speaking, the commonest form of arthritis, called osteoarthritis, is due more to wear and tear in the joints than to inflammation. In osteoarthritis, the cartilage which protects the ends of the bones wears away until the bones start to rub against each other. This is very painful. It also damages the bones, which rub away until the joint no longer fits together properly and may become inflamed.

Inflammation is the body's normal reaction to injury. White cells move to the site of injury and try to patch things up. The problem is that, sometimes, they overdo it. They set off a chain reaction, so that more and more cells turn up until the whole area becomes swollen and abnormal, and the process is hard to reverse. In some cases infection sets in and this makes things even worse.

Inflammation plays a big part in the less common form of joint damage - rheumatoid arthritis. This tends to occur earlier than osteoarthritis and gets worse more quickly.

Arthritis can affect any joint - hip, knee, spine, fingers, toes. If it happens in the hips or knees it can badly affect mobility. If the hands are affected, day-to-day jobs like writing, washing, cooking, even making a cup of tea, can become difficult.

Bone thinning is another serious problem of ageing, especially of middle-aged women after the menopause. The femur, wrist bones and vertebrae are especially prone to fracture. A broken femur in an elderly woman is a serious problem and can even be fatal. A long period of inactivity, sitting or lying down waiting for the bone to mend, may lead to poor circulation and blood clots which cause a stroke or heart attack.

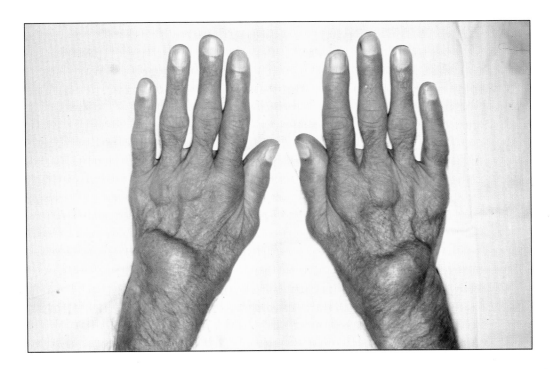

This woman's hands are swollen and painful with arthritis. The joints in her wrists and fingers are inflamed and damaged. Drugs can make her feel better and some of her joints can be replaced but as yet there is no cure.

This elderly woman used to be able to run as fast as you.

This metal hip joint has transformed life for someone with arthritis. He can now walk and run like anyone else and the joint should last for many years. Before his operation he could hardly get out of bed.

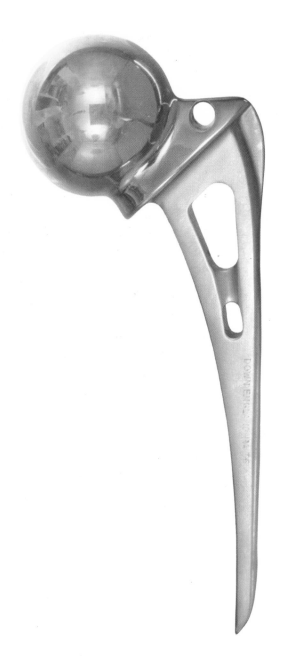

JOINT REPLACEMENT

Some older people have difficulty moving, not because their joints or bones get stiff or thin, but because the nerves that control their movements become diseased. At present very little is known about why this happens.

In Parkinson's disease, the chemicals that carry messages between nerves become faulty and the person gradually loses control of his or her movements. In motorneurone disease, the muscles become weak and cannot support the body properly. A lot of research is going into better treatment for these and other diseases associated with ageing, but at present there are no cures.

Today, many people are walking around with artificial joints in their hips, knees, shoulders, even their fingers and toes. The diseased joint is removed and replaced with an artificial metal or plastic implant.

The main problems are infection and loosening of the new joint. But better materials and great care in preventing infection during surgery are helping to reduce these problems.

IMMOBILITY

Someone who cannot move part of his or her body has some level of paralysis. This usually means that the nerves to the muscles in one area are so badly damaged that they cannot carry the messages needed to stimulate movement. The brain can still send instructions and the muscles may be capable of understanding them, but the vital communication link between the brain and the muscle is broken.

Paralysis can affect any or all parts of the body. In its mildest form it may affect a single muscle in the hand or foot. But paralysis can be much more widespread and severe. Someone who cannot move any of the lower half of the body, below the waist, is described as paraplegic. If there is no movement whatsoever in either the arms or the legs this is called quadriplegia.

Some people are born with some degree of paralysis. Parts of their bodies may have grown wrongly in the womb, or the body may have been damaged at birth. If too little oxygen gets to a baby's brain while it is being born, parts of the brain that control movement may be damaged. This is called cerebral palsy. Children with cerebral palsy have difficulty moving body parts the way they want to.

Some people become partially or totally paralysed as the result of an accident causing damage to the spine. If the vertebrae are broken or crushed they may be repaired. But if the spinal cord is broken, messages can no longer get to certain parts of the body. The amount of paralysis will depend on where the cord was broken. If it was broken in the lower part of the back then only the legs and lower body will be paralysed. But if the break is higher up, in the neck, the paralysis may be much greater.

It isn't only arm and leg muscles that are paralysed. Organs also need instructions carried by nerves. So paralysis below the waist can cause trouble with the bladder and intestines as well as the legs. Doctors have to pay careful attention to potential problems like urine infections.

This woman had a stroke which left her paraplegic – she is paralysed below the waist. She is learning how to look after herself again.

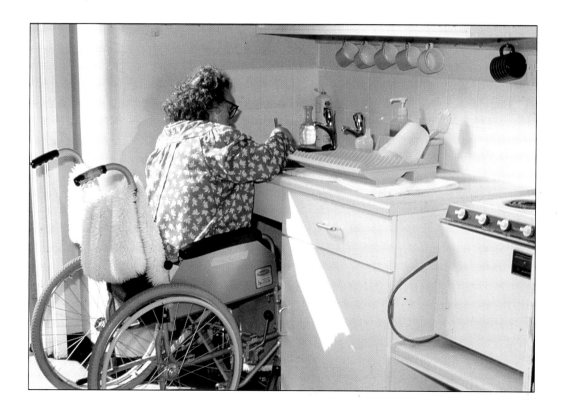

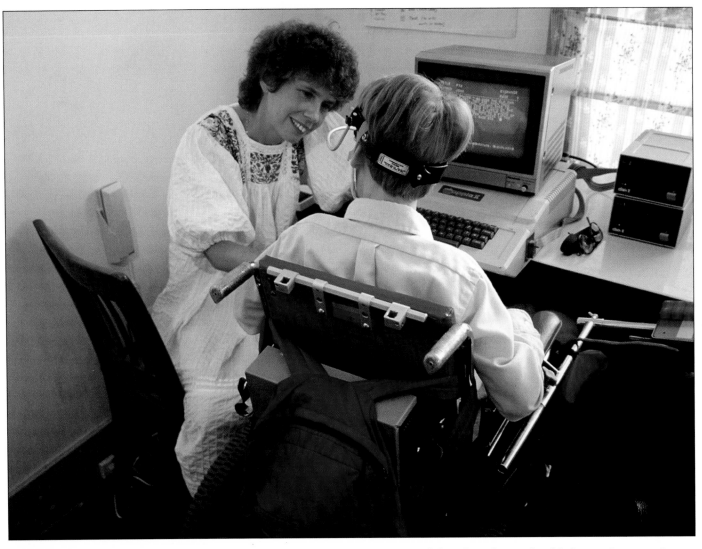

A computer technician shows a boy how to use a computer with a headwand which can be used instead of hands. This will allow him to work, write letters and 'talk' with friends.

MENDING DAMAGED BODIES

Until quite recently doctors thought that nerves didn't mend and that someone who was paralysed would never recover. Today, they know that some nerves do re-grow, so paralysed people may get some movement back. But no one can promise miracles. Someone who has lost all movement due to a broken neck is unlikely ever to walk again. However, often a lot can be done to get some movement back in hands or feet, so the injured person has some degree of independence and does not have to be looked after twenty-four hours a day.

Advances in computer technology have been a great help. Now, someone with only very limited finger movement may be able to operate a computer keyboard through which they can switch on lights and operate equipment such as TV, video, or hi-fi. Better still they can communicate with other people through computers, and access libraries and data banks.

At present, scientists are working on ways of bypassing damaged nerves. In the end, they hope to enable people to move by stimulating their muscles. Some progress has been made, but there is still a long way to go.

REPLACING LIMBS

Each year, a small number of babies are born with part or all of one or more of their limbs missing. Other people, children and adults, lose arms or legs in accidents or have to have them taken off because they are diseased.

Usually these people are offered artificial limbs. Each limb is carefully designed to fit the person who needs it and is strapped on to the stump that remains after an arm or leg is amputated.

The success of an artificial limb depends partly on how much of the limb is missing. Someone who has only had the lower part of one leg removed will do better, for example, than someone who has lost the leg from the thigh downwards. This is because it has not, as yet, proved possible to design an artificial limb with a satisfactory joint. So, without a knee joint, someone with an artificial leg walks very stiffly.

Scientists have begun to make battery-operated artificial limbs that allow a degree of mobility. Hands and arms are most successful. They use the patient's own remaining muscles to stimulate the artificial hand to open and close. These new limbs have proved most effective for children born with part of an arm missing. They seem to be more able to adapt to operating the new arm.

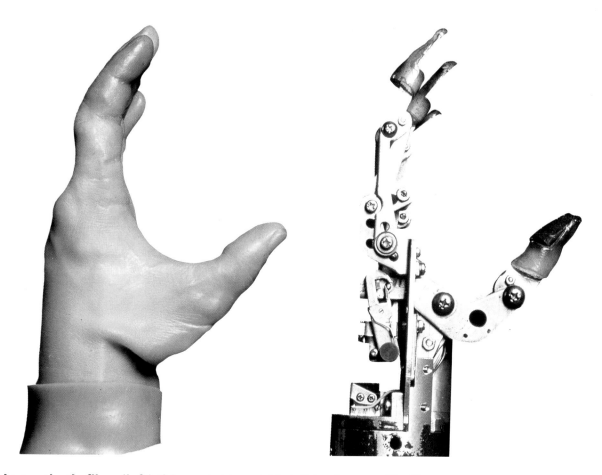

Covered in synthetic fibre (left) this metal hand (right) looks very life-like.

Scientists are currently researching ways of improving artificial limbs.

People who lose fingers and toes in accidents can often have them put back. Surgeons do these operations by looking down a type of binocular microscope, which magnifies everything so that they can see what they are doing. They use tiny scalpels and other instruments to join skin, blood vessels and connective tissue back together.

Although the fingers and toes look almost normal when they are put back, they do not always work very well. Unfortunately, it isn't possible to repair the nerves that have been cut and they do not always grow again. However, quite understandably, most people prefer to have fingers and toes in place, even if they do not work normally, than for them to be missing altogether.

Some attempts have been made to replace whole ams and legs, but these have been much less successful. It is almost impossible to join up all the blood vessels, so cells die and it may be necessary to remove the limb again.

The success of all operations to replace limbs depends on how quickly surgery can be done after the accident. It is very important to go straight to hospital, preferably with the severed finger or toe wrapped in ice.

AVOIDING AMPUTATION

Many children with bone cancer no longer need to have an affected arm or leg removed. This is thanks to a new operation in which the piece of bone with the tumour is removed and replaced with a small piece of bone from somewhere else in the child's body, or with donated bone. The new piece of bone fills the gap and links up with the original bone. Some surgeons use synthetic bone and can even fit a device that enables them to extend the new bone as the child grows.

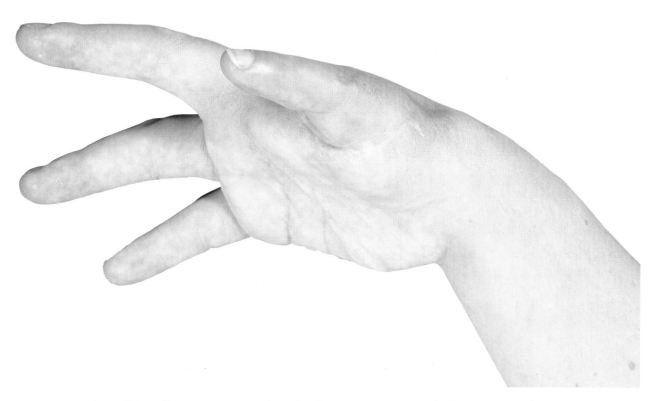

This girl was born with all her fingers but no thumb. Surgeons successfully moved a finger down to act as a thumb.

ANAESTHETICS

No one likes having an operation, but at least it's painless. Imagine what it must have been like to be cut open without an anaesthetic!

The first time an anaesthetic was used publicly was in the USA in 1846. The surgeon used ether during an operation to remove a tumour from a young man's jaw. The patient felt no pain.

For centuries doctors and dentists had used alcohol, herbs, hypnosis or a sharp bang on the head to put their patients to sleep while they operated on them. But we can thank Queen Victoria for making it acceptable to have an anaesthetic. In 1853 she was given chloroform to relieve her pain during the birth of her son, Prince Leopold.

Strictly speaking, an anaesthetic is a drug that takes away all feeling, while a drug that just gets rid of the pain is an analgesic. But a lot depends on how much of a drug you give and for how long. For example, some drugs make you feel relaxed in small doses but send you to sleep when given in larger amounts.

An anaesthetic that puts you to sleep is called a general anaesthetic. Some, like ether, chloroform and nitrous oxide (laughing gas) are gases that are breathed in. Others are given by injection. Often, people are also given a muscle relaxant, so that surgeons do not have to fight against powerful muscles to get at the internal organs. But these drugs relax the muscles in the airways as well as the big muscles in the chest and abdomen, so patients need a machine to help them to breathe.

Many of the anaesthetics that are used today are much shorter acting than the drugs used even twenty years ago. This means that patients wake up much more quickly after an operation.

Many operations are done under local anaesthetic too. 'Local' means that only the part of the body needing treatment is anaesthetized and the patient stays awake throughout the operation. This is especially useful for elderly and overweight people and for those with breathing problems who might react badly to a general anaesthetic.

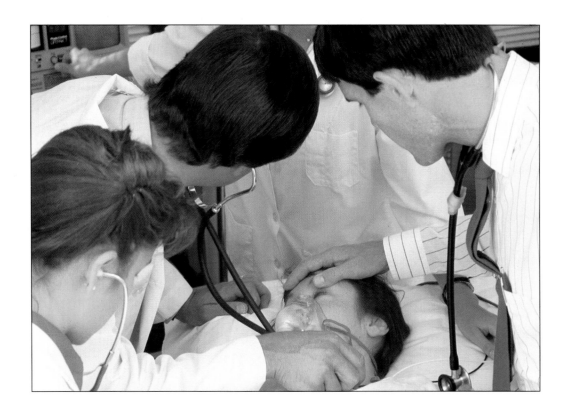

When he wakes up, this little boy won't remember anything about his operation. The anaesthetic he is given will make sure he is asleep while surgeons take out his appendix.

THE PAIN BARRIER

ABOVE This man should be in great pain – those coals are very hot. But he has trained himself to ignore the messages coming from the nerves in his feet. Please don't try this yourself! You would burn your feet very badly indeed.

Some people can put up with a lot more pain than others. This means that they can be in more pain before their nerves tell their brain that something hurts.

Lots of things can affect how well you cope with pain. You've probably heard about soldiers wounded in battle who fight on without realizing how badly hurt they are. They know their lives depend on keeping moving and their brain blocks out the pain.

Compare that with someone who is worried about losing his or her job or someone who is unhappy at home. When a person is tense and anxious, even a slight headache or stomach ache may feel very painful.

It is usually possible to get rid of pain with painkilling drugs. But sometimes the pain won't go away until the sufferer's other problems are tackled. Their doctor may refer them for therapy with a trained counsellor.

Ellen Van Langen pushed herself through the pain barrier to win a gold medal in the 800m at the Barcelona Olympics, 1992.

SPORTS INJURIES

Rugby, hockey and climbing are the riskiest sports you can do. In the UK, the injury rate for these sports is higher than for many other activities, such as motor sports, horse riding and boxing, which sound more dangerous.

But that doesn't take account of how bad the injuries are. Your chances of escaping serious injury in a big pile-up while motor-racing must be less than if you find yourself at the bottom of a scrum in rugby. On the football pitch or running track you are likely to pull a muscle, tear a ligament or, at worst, break a bone, but boxing could possibly leave you with permanent brain damage or you might even die.

Sprains and strains are the most common sports injuries. They occur at body joints - when you are moving fast and reacting quickly, you can easily turn a limb at an awkward angle. If you sprain your ankle it means that you have twisted the joint and probably torn some of the ligaments that hold it together. A strain is similar but not usually so serious. Either way, your ankle will be painful and probably swollen. Ice reduces the swelling, and if you bandage the joint this helps to keep it still while it heals. You should rest it and not play any sport until it is better.

You can 'pull a muscle' by moving very awkwardly, especially if you haven't warmed up properly. The symptons - pain and swelling - are similar to those of a strain or sprain, but it is the muscle rather than the joints that hurts.

Your muscles will ache if you play a sport after you haven't exercised for some time, or if you overdo it. You may also feel stiff the next day. This is because your muscles have worked harder than usual. Heat may help to relieve the stiffness.

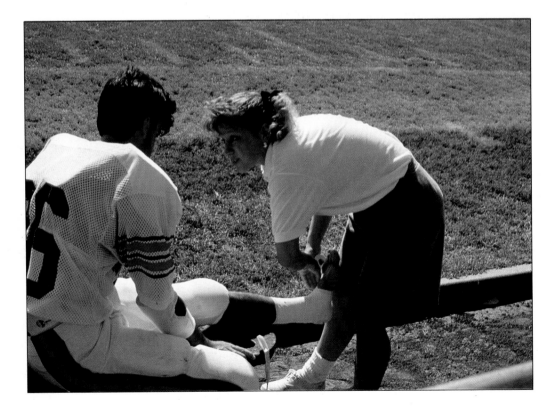

This footballer has hurt his ankle. It will be a while before he will be allowed to play again.

It's important to keep a broken leg completely still for several weeks.

Shin splints are another common problem for athletes. Sufferers find their shins hurt during or after exercise. This may be due to inflammation or swelling of the muscles. Resting and applying ice can help. Sometimes pain in the legs and feet is due to stress fractures. These are hairline cracks in the bones which may not show up at first on an X-ray. If you take more exercise they will get worse. Once again, rest is the answer. If the problem is ignored it only gets worse.

A broken bone will show up on an X-ray straight away and it must be set so that the bone grows straight when it mends. With some fractures, the broken limb is encased in plaster. But if there are several breaks or the bone is in a difficult place – just above the elbow, for example – it may be better to put a metal pin through the bone to hold it firm while it mends.

TOO MUCH EXERCISE TOO YOUNG

Top tennis player Monica Seles had to drop out of several tournaments in 1991 because of injuries. And sadly other young players, including Tracey Austin, have had to give up altogether because of injuries. It may be that some talented young athletes are training and playing more often than is good for them.

Teenage athletes are still growing and developing and their bones and muscles are not as tough as those of an adult. So some young athletes may have to stop training so hard.

AVOIDING INJURY

If you decide to take up a sport it is very important that you learn how to do it correctly so that you don't injure yourself. That means taking it slowly and learning the basics before trying to do anything more adventurous.

Most sports need fitness and this only comes with training. You cannot expect to go and kick a ball round a football pitch for an hour and become fit. You should build up your fitness by gradually increasing the amount of time you spend doing your chosen sport. You'll need to stick with it too!

Never start exercising without warming up. By warming up for ten or fifteen minutes you gently stretch muscles and ligaments before you make them start to work hard. A warm up also helps to get your blood circulating well. Small blood vessels to your muscles and skin start to open up, supplying extra oxygen and nutrients.

Warm up each part of your body in turn, starting at the top. Roll your head around on your shoulders clockwise and then anti-clockwise. Relax your shoulders by moving them gently up and down.

Stretch your arms out in front of you. Swing them up and down and then slowly around each shoulder, like a windmill action. Next, swing your arms from side to side so your body moves with your arms from the waist up, but keep your feet flat on the floor and your knees slightly bent.

Now, warm up your legs. Stand straight and bend forwards, keeping your back straight, until you can feel the backs of your legs pull slightly. Don't overdo it! Next, kneel down on the floor and stretch out each leg alternately behind you. Finish off with a little gentle running on the spot.

Some sports need more specific warm up exercises and there are many others to choose

head rolls

arm stretches

GETTING KITTED OUT

Getting the right equipment is just as important as warming up properly. For most sports your most important piece of equipment will be your shoes. When you are running and jumping, your feet and lower legs absorb a lot of impact. It is much better to spend your money on a good pair of sports shoes that support your feet and ankles than to splash out on designer shorts or shirts. Clothes fashions come and go but you only get one pair of feet!

Your feet should never move about inside your shoes. So whatever brand-name and style you choose, be sure the shoes fit you correctly and that you lace them up firmly.

from so that you don't get bored. After your warm up, go through a series of drills to loosen you up. Most sports have practice drills to get you moving and help with co-ordination. That's why it's worth going to a class where the teacher or coach can explain them to you.

Some types of exercise need a 'cool down' as well as a warm up. You shouldn't just stop, for example, at the end of an aerobics class. After the most tiring exercises you should relax back into an easier routine and let your heart rate gradually fall and your muscles ease off.

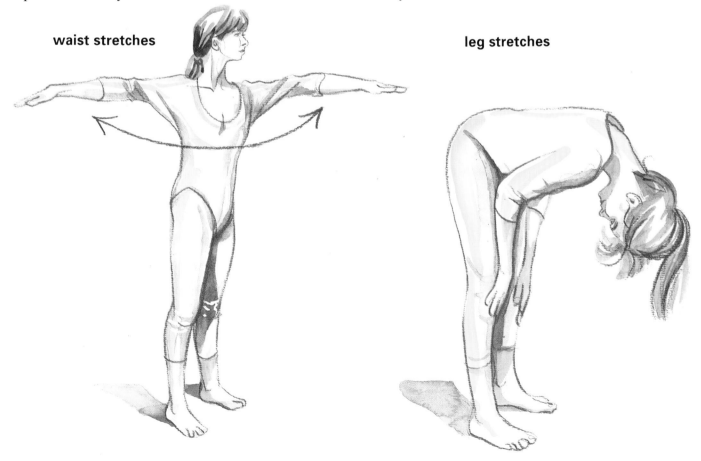

waist stretches

leg stretches

FULL SPEED AHEAD

On 25 August, 1991 the American sprinter Carl Lewis ran the 100 m faster than anyone had ever done it before. His time was 9.86 seconds. It was a remarkable achievement, but it was less than half a second faster than the world record set for the same event in 1936 by Jesse Owens. Today's athletes can run faster than their counterparts did over half a century ago, but a top athlete like Lewis can only expect to improve his best running time by about 0.2 seconds during his entire career.

The cheetah is the fastest land animal and it can move at speeds up to 105 kilometres per hour (kph). Greyhounds and racehorses can smash the 64 kph mark. But male sprinters can only reach speeds of about 35 kph. What is it that limits how fast a man or woman can run?

Quite simply, humans aren't built for speed. You may have noticed already that the fastest animals on earth have four legs, not two! Their muscle power is much greater and they have a higher proportion of the 'fast twitch' muscle fibres that are needed for speed.

Athletes of the future will probably never go very much faster than they do today. Better running tracks and lighter, higher performance shoes will help them to improve their times a little. But this will be a result of technology, not of better training or greater muscle power.

LEFT Carl Lewis – one of the fastest men the world has ever seen.

RIGHT Florence Griffith-Joyner (centre) won the women's 100m at the Seoul Olympics in 1988. But she wouldn't stand a chance against a greyhound or a racehorse!

BREEDING WINNERS

Athletic parents tend to have athletic children. That's a sweeping statement and you probably know someone who's a demon on the football pitch but whose parents can barely kick a ball. Training, enthusiasm and the will to win can go a long way to overcome any athletic handicaps which we inherit from our parents. But you only have to look at some sporting families - the footballing Charltons, the tennis-playing Sanchez clan and the cricketing Cowdreys - to realize that it helps to have some athletic genes.

No one has yet worked out what genes you need to make you a sprinter, a swimmer or a squash player. But scientists all over the world are working on it. They are spending a lot of time and money investigating the human genome. This is the blue print of genes which we carry in every cell of our bodies. It determines what we look like and how we feel.

There are 50,000 genes in the human genome and, so far, scientists have discovered what about 2,500 of them do. They are also beginning to understand more about what turns them on and off. A few doctors have even tried to correct abnormal genes that put some people at risk from life-threatening diseases. The science of altering the natural characterstics of a person's genes is called genetic engineering.

In theory, therefore, it could be possible for us to breed super-athletes. It would probably take another fifty years. But scientists might be able to use genetic engineering to improve people's athletic abilities.

Will they do it? Probably not. Scientists are well aware of the ethical problems of changing our genes. Genetic engineering may be justifiable to save someone's life or to prevent them being born with a disabling muscle or nerve disorder. But is it right to breed people with certain physical or mental characteristics? Society will have to decide how far it wants to go in pursuit of the perfect mind and body.

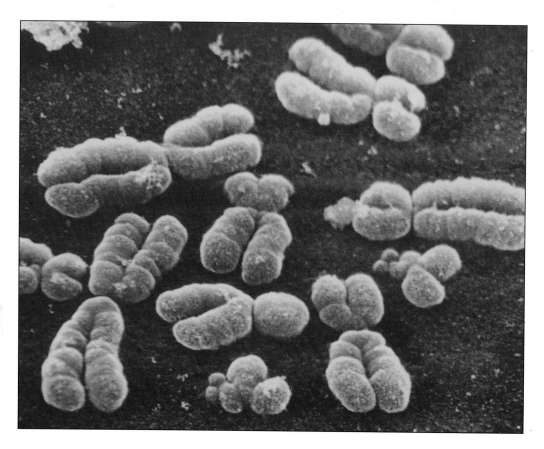

Some of the chromosomes – magnified many times – found in each of our cells. They carry the information that helps decide whether someone becomes an Olympic athlete or not.

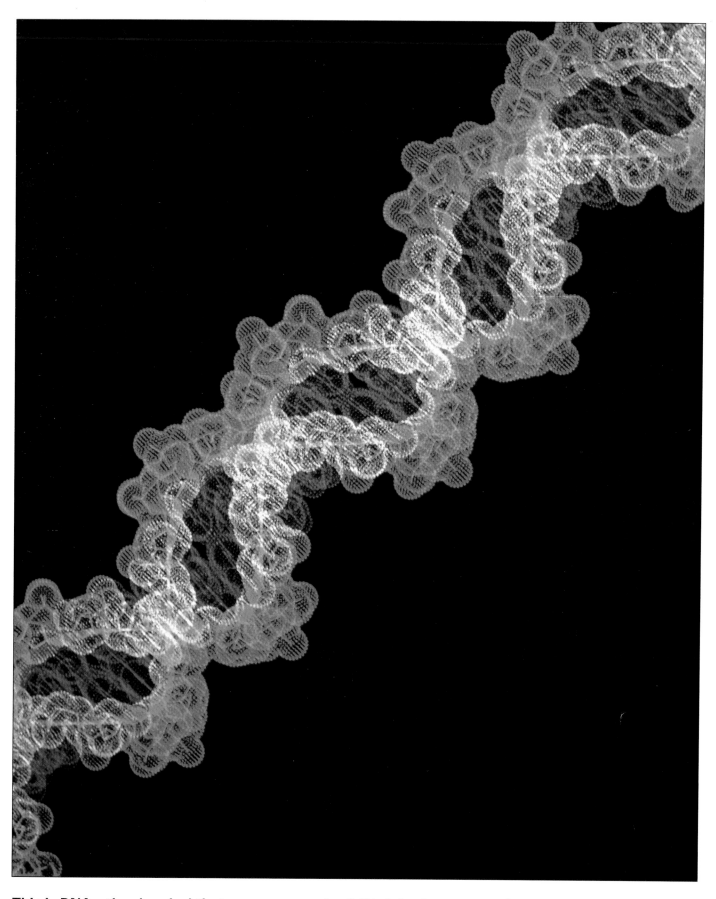

This is DNA – the chemical that genes are made of. We inherit our genes from our parents.

BODY LANGUAGE

There's more to movement than just getting about. The way we move says a lot about how we feel and how we want other people to think about us. Put pop singer Madonna on stage and she oozes sex appeal. It's not just her outrageous costumes. Every move she makes has been carefully chosen to make her audience like her act.

We all want to be liked. That's human nature. When we meet someone new, we smile and shake hands. These are welcoming gestures. When we see someone we like, we smile and wave and try to catch their eye. But if we see someone we don't like, we keep our eyes to the ground and walk past as if we have not seen them.

This is fairly obvious body language. Sometimes we need to use much more subtle movements to show how we feel. We may want to challenge what someone is saying without actually telling them we don't believe them. Or we may want to convince them we are right when we aren't at all sure of our facts!

Watch politicians being interviewed on television and you'll see this sort of body language! Ideally they will want to speak from behind a desk so they can lean forward on it and look confident. This says 'I'm honest and confident; you can believe what I'm saying.' They hate being interviewed in a low chair because they will look small, uncomfortable and unconfident. They mustn't lean back too far or they will look as though they are frightened of the interviewer.

Look at their eyes when they are answering questions. They try always to look straight at the interviewer so that they look truthful. We do the same when we are talking to someone. If someone is lying they find it hard to look you in the eye.

In a political interview, the interviewer's body language is almost as interesting as that of his subject. If he is challenging the politician he will sit upright or lean slightly forward and hold eye contact. If he doesn't believe his subject he may sit

back slightly, rest his chin on his fingers and look disbelieving. If he is bored he will slouch in his chair, lean on one arm and possibly even yawn! Such tactics are all geared to put his subject off. Then, when he senses a break in the defences, he will suddenly lean forward and go for 'the kill'.

RIGHT American politician Jesse Jackson uses body language to stress a point. By raising his right hand and pointing, he grabs the attention of the 'audience' and makes his speech all the more persuasive.

LEFT Pop star Madonna owes her success to expert use of body language as well as her voice.

LISTENING TO BODY LANGUAGE

Your new friend tells you his parents are millionaires. Should you believe him? Watch his body language and ask yourself the following questions:

● Does he a) look you in the eye when he tells you this or b) look around the room?
● Is he a) standing still or b) shifting from one foot to the other?
● Are his hands a) relaxed or b) fiddling with something?
● Is he a) prepared to stay and talk about how his parents became millionaires or b) in a big rush to go and do something else?

If the answers are all 'b' take what he says with a pinch of salt! If the answers are all 'a' find out where his parents keep their money!

STAYING MOBILE

More of us can expect to live into our seventies and eighties than at any time in history, but living longer has its drawbacks. Elderly people aren't as mobile as when they were younger. Most have some arthritis, so their joints are stiff. Sitting down a lot makes their muscles weak and a lot of elderly women have thin bones which may break. How likely we are to get arthritis and other ageing diseases depends a little on the genes we have inherited from our parents.

We can't stop the ageing process but we can at least delay its effects. Eating sensibly and staying fit when we are young can help us to stay mobile as we get older. For example, if you can keep at roughly the right weight for your height you will reduce the strain on all parts of your body, including your heart, lungs, joints and bones.

You will also feel more inclined to take regular exercise – you probably don't know anyone who is very overweight and likes to run!

Regular exercise keeps your joints mobile and your muscles strong. It also gets you into a routine which can last you all your life. If in your teens you get used to playing football, squash or tennis, or going to a gym every week, you'll probably go on doing it when you leave school.

Exercise should be a pleasure, not a punishment! So choose a sport that you enjoy. Unfortunately, a lot of people are put off sports at school because they aren't very good at the team games that are played. You may prefer swimming, badminton or running. All these activities are good for your body and will help keep your bones, joints and muscles in shape for when you are older.

Exercise is good for you, whatever your age.

A well-balanced diet is also important. No more than a third of your energy intake (calories) should come from fatty foods. The rest should come from protein and carbohydrate. You also need fruit and vegetables for fibre, vitamins and minerals to keep you in peak condition.

Your bones may be strong now, but as you get older they will start to get thinner and weaker. That's why it's important to build strong bones while you are young, and for that you need plenty of calcium. You get calcium from fish, milk, cheese, yoghurt and nuts.

Drugs can help to keep bones strong in middle age, but they can't make up lost time if you've neglected your body. There are also some problems directly related to growing older. Many women are advised to have hormone replacement therapy (HRT) after the menopause to put back the female hormones that they are no longer making for themselves. This is because, without these hormones, their bones may get brittle.

There is no magic potion to keep us young. In the end our bones and joints will wear out. But you'd probably prefer that to happen when you're very old, not in your sixties. It's up to you!

ABOVE Cheese contains calcium – the mineral you need to build strong healthy bones. It is also delicious and it comes in a huge number of varieties. What is your favourite cheese?

BELOW Ben Johnson, the disgraced Olympic gold medal winner who took anabolic steroids to improve his running. These drugs help to build muscle but can have extremely dangerous side effects.

ANABOLIC STEROIDS

The Canadian athlete Ben Johnson was thrown out of the 1988 Olympic Games in Seoul because he took drugs to improve his performance. The drugs which athletes are most often tempted to take are called anabolic steroids. They help to build muscle but they have serious side effects and can cause strokes and heart attacks. If women take them they have a masculinizing effect – the women's skin becomes thicker and coarser and they develop more body hair.

Any athlete who is found to have taken drugs will be disqualified from a competition and usually banned from competing for several years – Ben Johnson has only recently been reinstated. Anabolic steroids are not even particularly effective at improving performance. The results are very variable and hard to predict, so it is just not worth using them.

GLOSSARY

Actin a protein found in muscles which is needed for muscle contraction and relaxation.

Aerobic processes occurring within the body which need oxygen.

Anaerobic processes within the body which do not need oxygen.

Arthritis swelling and/or weakening of joints.

Biceps the muscle found at the front of the upper arm.

Cartilage tough gristly material covering parts of the bones; it acts as a shock absorber in joints.

Gene the basic unit of living tissue; in humans, an individual's genes are inherited from the parents and determine physical or mental characteristics.

Glucose a chemical found in many foods that gives energy.

Glycogen a chemical used by muscle to produce energy.

Isometric the description of a muscle contraction in which the muscle does not shorten but uses its energy to fight gravity; for example, when you hold a weight above your head.

Isotonic the description of a muscle contraction in which the muscle shortens in order to move part of the body.

Joint the point at which two bones meet.

Ligament a band of fibrous tissue which ties bones together.

Menopause the time, usually around fifty years of age, when a woman stops producing female hormones and can no longer become pregnant.

Myosin a second protein found in muscles and needed for muscle contraction.

Nerve a hair-like structure that carries messages from the brain and spinal cord to a different part of the body.

Osteoporosis bone thinning, which occurs after the menopause and can lead to fractures.

Paraplegia paralysis of the lower part of the body.

Quadriceps the large muscle in the thigh.

Quadriplegia paralysis from the neck downwards.

Reflex involuntary or automatic muscle response to stimulation.

Tendon fibrous material which binds muscles to bones.

Triceps muscle in the back of the upper arm, which contracts when the biceps relaxes and vice versa.

Vertebrae bones that make up the spine, or backbone.

BOOKS TO READ

The Human Machine by Brenda Walpole (Wayland, 1990)

The Body and How it Works by Steve Parker (Dorling Kindersley, 1987)

The Don't Spoil your Body Book by Claire Rayner (Bodley Head, 1989)

The Human Body by Ruth and Bertel Bruun (Kingfisher Books, 1985)

Let's Discuss Health and Fitness by Tony Wheatley (Wayland, 1988)

Disease and Discovery by Eva Bailey (Batsford, 1985)

Eat Well by Miriam Moss (Wayland, 1992)

Keep Fit by Miriam Moss (Wayland, 1992)

Pocket Book of the Human Body by Brenda Walpole (Kingfisher Books, 1987)

Twentieth Century Medicine by Jenny Bryan (Wayland, 1988)

ACKNOWLEDGEMENTS

Action Plus 39 (Tony Henshaw);All Sport UK cover and title page (Gray Mortimore), 38 (David Cannon), 45 (Gray Mortimore); Chapel Studios 23 (bottom), 42; Eye Ubiquitous 4; Hutchison Library 33 (top); Life Science Images cover background;Medical Illustration: Institute of Child Health & The Hospitals for Sick Children 31; National Medical Slide Bank 27; Rex Features 42,43 (top, Estrin); Science Photo Library 9 (top, Eric Grave), 17 (NASA), 28 (Lowell Georgia), 29 (Claude Charlier), 33 (bottom, Rosal), 40 (Biophoto Associates), 41 (Will & Deni McIntyre), 44 (Hank Morgan); Skjold Photographs 20 (bottom), 34; Tony Stone 8, 43 (bottom); Topham Picture Library 9 (bottom), 10, 18, 20 (top), 22, 23 (top); Wayland Picture Library 21, 45 (top); ZEFA 5, 6, 7 (right), 11, 12, 16, 19, 24, 25, 26, 30, 32, 35, 37 (top), 39, 43. Artwork: Debbie Hinks

INDEX

anaesthetics 32
arthritis 24, 44
  osteoarthritis 24
  rheumatoid 24

balance 6
bones 4, 6, 16, 18, 21, 24, 35, 44
  broken 35
  femur 6, 24
  fibula 6
  metatarsals 6
  pelvis 6
  phalanges 6
  spine (see spine)
  tarsals 6
  tibia 6
brain 6, 8, 20, 23, 27, 33

calcium 16
calories 8
carbohydrate 8, 10, 12
cartilage 9, 24
cerebral palsy 27
circulation 24, 36
co-ordination 18, 19, 20
cramp 12

energy 8, 10, 12
exercise 24, 36, 37, 44
eyes 18

fat 8, 12
feet 6, 8, 27
fingers 9
fitness 4, 13, 36, 44
flying 42
food 8, 10, 12, 44

genes 40, 46
glycogen 10, 12, 46
gristle 9

hands 15, 18, 27
  left 20
  right 20
hips 4, 6, 9, 24
'hot housing' 19

injuries 23, 27, 28, 31, 33, 34, 35, 36, 37
  shin splints 35
  sprains 34
  strains 34
isometric 13
isotonic 13

Johnson, Ben 45
joints 4, 6, 9, 16, 18, 24, 29, 34, 44
  ball and socket 9
  gliding 9
  hinge 9
  replacement 26
  saddle 9

knees 4, 6, 8, 9, 21, 24

Lewis, Carl 38
lifting 13
ligaments 4, 34, 36
limbs 4, 6, 8, 9, 15, 29
  artificial 29

motor neurone disease 26
muscles 4, 6, 8, 10, 13, 15, 16, 21, 23, 24, 27, 29, 34, 36, 44, 45
  biceps 15
  cells 8
  extensor 8
  fast 10, 12
  fibres 8, 12
  slow 10
  flexor 8
  hamstring 6

pectoral 42
skeletal 4
smooth 4

nerves 4, 8, 18, 20, 21, 26, 28, 33
nerve transmitters 8
nervous system 21

Owens, Jesse 38
oxygen 10, 27

paralysis 27, 28
Parkinson's disease 26
proteins
  actin 8
  myosin 8

quadriplegia 27

reflexes 21, 22

shivering 23
shoulders 9
skeleton 16
spine 6, 24, 27
  discs 6, 9
  spinal cord 6, 8, 21
  vertebrae 6, 24
sport 10, 12, 13, 16, 18, 22, 24, 26, 34, 35, 36, 38, 40, 44
stamina 10, 12, 13
strength 13

tendons 4, 21
thighs 6
tremors 23
twitches 23

walking 8, 13, 18
wrists 9
writing 18, 19, 20